My First Animals
Copyright © 1990 by Bettina Paterson
First published in 1990 by William Heinemann, London, England,
under the title *Animal Friends*.

Library of Congress Cataloging-in-Publication Data
Paterson, Bettina.
  My first animals / Bettina Paterson.
    p.   cm.
  Summary: Illustrations consisting of paper cutouts introduce well-
known animals such as the frog, duck, and pig.
  ISBN 0-690-04775-4 : $. - ISBN 0-690-04777-0 (lib. bdg.):
$
  I. Animals-Pictorial works-Juvenile literature. [I. Animals-
Pictorial works.] I. Title.
QL49.P29 1990   89-17275
59l-dc20   CIP
                                                          AC

Printed in Belgium. All rights reserved.
1 2 3 4 5 6 7 8 9 10
First American Edition, 1990

My First Animals

Bettina Paterson

Thomas Y. Crowell   New York

duck

frog

COW

goat

squirrel

robin

moose

goose

rooster

fox

skunk

pig

owl

hen

donkey

sheep

kingfisher

horse

otter

badger

deer